Abdelhafid Mimouni

Lysine Decrypted

Abdelhafid Mimouni

Lysine Decrypted

ScienciaScripts

Imprint

Cover image: www.ingimage.com

This book is a translation from the original published under ISBN 978-620-6-72017-1.

Publisher:
Sciencia Scripts
is a trademark of
Dodo Books Indian Ocean Ltd. and OmniScriptum S.R.L publishing group

120 High Road, East Finchley, London, N2 9ED, United Kingdom
Str. Armeneasca 28/1, office 1, Chisinau MD-2012, Republic of Moldova, Europe
Printed at: see last page
ISBN: 978-620-8-20211-8

Lysine Decrypted

Author:

Dr. Abdelhafid Mimouni independent researcher specialized in bioinorganic systems chemistry. Extensive expertise in macromolecular synthesis and characterization. He obtained his PhD in chemistry from the University of Paris XII in 1997 and a Diplôme des études approfondies in bioinorganic systems from the University of Paris XI in 1993.

Foreword

This book, **"Lysine Decrypted"**, explores the crucial importance of this essential amino acid for our well-being. Often overlooked, lysine plays a central role in protein synthesis, bone health, collagen formation and immune system support. By providing detailed information on its biological roles, dietary sources, and the consequences of its deficiency, this guide aims to enlighten readers on how lysine can contribute to better health. Through practical advice and recent research, we hope to help everyone effectively integrate this vital amino acid into their daily diet and reap its many health benefits.

TABLE OF CONTENTS

Introduction

Introduction to Lysine: Definition, Importance and Overview

Definition of Lysine

Lysine is an essential amino acid, which means it must be obtained from the diet, as the human body cannot synthesize it. It is one of nine essential amino acids for adults and one of ten for children. As a key component of proteins, lysine plays a crucial role in many biological processes.

Importance of Lysine

Lysine is vital for several bodily functions, including :

- **Protein synthesis**: It is a fundamental constituent of proteins, contributing to the formation and repair of body tissues. Lysine is essential for growth, muscle repair and the production of enzymes and hormones.
- **Calcium absorption**: Lysine improves calcium absorption in the intestine, contributing to healthy bones and teeth.
- **Collagen formation**: necessary for the synthesis of collagen, a key protein in connective tissues such as skin, tendons and ligaments.
- **Carnitine production**: Lysine is involved in the production of carnitine, which helps convert fatty acids into energy.
- **Immune function**: Supports the immune system by promoting antibody production and strengthening the body's natural defenses.

Overview

Lysine is mainly found in animal proteins such as meat, fish and dairy products. However, it is also present in some plant sources, albeit often in lesser quantities. Because of its importance in various bodily functions, it is crucial to ensure an adequate supply of lysine through the diet or, if necessary, supplements.

Why Study Lysine? Book objectives

Objectives of the Lysine Study

The in-depth study of lysine is essential for several reasons:

- **Understanding Nutritional Requirements**: Understanding how lysine contributes to health enables better diet planning to prevent deficiencies and optimize overall health.
- **Deficiency management**: Identifying symptoms of lysine deficiency and groups at risk helps to implement appropriate nutritional strategies to avoid health problems linked to a deficiency of this amino acid.
- **Medical applications**: Knowing about lysine's medical applications, such as its use in the treatment of herpes infections, provides a better understanding of its potential therapeutic benefits.
- **Optimizing Food Sources**: Learning how to effectively integrate lysine into the daily diet helps maximize its benefits while diversifying nutritional sources.

Book objectives

The book aims to :

1. **Educate on Lysine**: Provide a clear and detailed definition of lysine, its biological importance, and the roles it plays in the human body.
2. **Explore Biological Roles**: Detailing the various biological functions of lysine, including protein synthesis, calcium absorption, collagen formation and more.
3. **Identifying Dietary Sources**: Present the main dietary sources of lysine, both animal and vegetable, and give practical advice on how to incorporate these sources into the daily diet.
4. **Dealing with deficiencies and risks**: Describe the causes and consequences of lysine deficiency, as well as the groups of people particularly at risk.
5. **Discuss Medical Applications and Supplementation**: Examine the medical uses of lysine, the different forms of supplement available, and the results of clinical studies.
6. **Offer Tips for a Balanced Diet**: Provide practical advice on planning lysine-rich meals and easy-to-prepare recipes.

This book aims to provide comprehensive and accessible information on lysine, enabling readers to better understand this essential amino acid and integrate it optimally into their diet and daily lives.

Chapter 1: Understanding Lysine

Definition and Structure

Definition of Lysine

Lysine is an essential amino acid, which means it cannot be synthesized by the body and must therefore be obtained from the diet. It is one of the twenty amino acids that make up proteins in the human body. Amino acids are the basic units of protein, and each amino acid plays a specific role in protein formation and function.

Chemical structure

Lysine is represented by the chemical formula $C_6H_{14}N_2O_2$. Its chemical structure is characterized by the presence of :

- **An Amino Group** (-NH_2): This group is located at the end of the carbon chain, making lysine basic.
- **A Carboxyl Group** (-COOH): This group is present at the other end of the chain, giving lysine its acidic properties.
- **An Amino Side Chain**: Lysine has an amine side chain (-NH_2) located at the fourth position of the central carbon. This side chain is what distinguishes lysine from other amino acids.

Visually, the chemical structure of lysine is often represented as follows:

H2N

|

H2N-C-COOH

|

CH2

|

CH2

|

CH2

|

NH2

Essential and Metabolism

Why is Lysine Essential?

Lysine is classified as an essential amino acid because the human body cannot synthesize it endogenously. It must therefore be ingested through the diet. Without sufficient lysine, various biological processes, including protein synthesis, growth and tissue repair, cannot proceed properly. Lysine deficiency can lead to growth problems, fatigue and metabolic disorders.

Lysine metabolism

Lysine metabolism involves several key steps:

1. **Intestinal absorption** :
   - After ingestion, lysine is absorbed in the small intestine. It is transported by specific transporters into the intestinal cell membrane and then released into the bloodstream.
2. **Transport and Distribution** :
   - Once in the blood, lysine is transported to various tissues and organs, where it is used for protein synthesis and other biological functions. It is also stored in muscle and other body tissues.
3. **Use and conversion** :
   - In cells, lysine is incorporated into proteins during messenger RNA (mRNA) translation. It plays a crucial role in the formation of collagen and other structural proteins.
   - In addition to its role in protein synthesis, lysine is converted into various metabolites. It can be converted into acetyl-CoA, a key component of the Krebs cycle, which contributes to energy production.
4. **Excretion** :
   - Excess lysine and its metabolites are excreted mainly by the kidneys in the urine. Regulation of lysine concentration in the body is therefore closely linked to kidney function.

Chapter 2: Biological roles of Lysine

Protein synthesis

Explaining the Role of Lysine as a Protein Building Block

Lysine plays a fundamental role in protein synthesis as an essential building block. Proteins are made up of long chains of amino acids that fold into specific three-dimensional structures to perform various biological functions. Lysine is integrated into these polypeptide chains and is crucial for the formation of functional proteins in the body.

- **Integration into polypeptide chains**: During translation of messenger RNA (mRNA) into proteins, lysine is added to polypeptide chains according to the genetic code. Using mRNA as a template, ribosomes assemble amino acids in the specified order to form proteins.
- **Importance in muscle development and tissue repair**: Lysine is particularly important for muscle development and body tissue repair. It contributes to the formation of muscle proteins such as myosin and actin, which are essential for muscle contraction and growth. In addition, during wound healing, lysine helps rebuild damaged tissue by providing the amino acids needed for repair protein synthesis.
- **Enzyme and hormone production**: Lysine is involved in the production of various enzymes and hormones. Enzymes, which are specialized proteins, catalyze vital chemical reactions in the body. Hormones, also

proteins, regulate various physiological processes, such as growth, metabolism and reproduction.

Calcium absorption

How Lysine Improves Calcium Absorption in the Gut

Lysine helps improve calcium absorption in the gastrointestinal tract through several mechanisms:

- **Increased Calcium Solubility**: Lysine helps maintain calcium in a soluble form in the intestine, facilitating its absorption by intestinal cells.
- **Interaction with calcium transporters**: lysine can influence the activity of calcium transporters in intestinal cells, thereby increasing the efficiency of calcium absorption.

Consequences for Bone and Teeth Health

- **Development and maintenance of bone health**: By improving calcium absorption, lysine plays a crucial role in the development and maintenance of bone density. Calcium is a major component of bones and teeth, and adequate absorption is essential to prevent conditions such as osteoporosis.
- **Calcium binding in bones**: Lysine helps bind calcium in the bone matrix, contributing to bone strength and resistance. This reduces the risk of fractures and promotes long-term bone health.

Collagen formation

Role in Collagen Synthesis

Collagen is the most abundant protein in the human body, making up around 30% of total proteins. Lysine is a key component in collagen formation:

- **Lysine hydroxylation**: In the collagen biosynthesis process, lysine undergoes a chemical modification called hydroxylation, which is essential for stabilizing the collagen structure. This step is catalyzed by the enzyme lysyl hydroxylase.

Impact on Conjunctive Tissue

- **Skin**: Collagen provides skin with elasticity and firmness. Adequate collagen synthesis, facilitated by lysine, helps maintain healthy, wrinkle-resistant skin.
- **Tendons and Ligaments**: Collagen forms the structure of tendons and ligaments, providing strength and flexibility. Sufficient collagen is essential for the repair and resilience of these connective tissues.
- **Blood vessels**: Collagen also contributes to the structure of blood vessels, helping to maintain their integrity and function.

Carnitine production

How Lysine Contributes to Carnitine Production

Lysine is necessary for the synthesis of carnitine, an essential molecule for fat metabolism:

- **Carnitine formation**: Lysine, in combination with methionine (another amino acid), is converted to carnitine in the body. This conversion takes place mainly in the liver and kidneys.

Carnitine's role in fat metabolism and energy production

- **Fatty acid transport**: Carnitine transports fatty acids across the mitochondrial membrane, where they are oxidized to produce energy. This process is crucial for energy metabolism, particularly during prolonged physical exercise or periods of fasting.
- **Energy production**: By facilitating the oxidation of fatty acids, carnitine contributes to the production of ATP, the main source of energy for cells.

Immune function

Effects of Lysine on the Immune System

Lysine plays an important role in modulating the immune response:

- **Immune Function Support**: Lysine helps strengthen the body's natural defenses by supporting the production of various immune system components, including antibodies.

- **Inhibition of Viral Infections**: Lysine can inhibit the replication of certain viruses, such as herpes simplex virus, by reducing their ability to multiply.

Contribution to Antibody Production and Global Immune Response

- **Antibody production**: Lysine contributes to the synthesis of antibodies, which are specialized proteins produced by the immune system to identify and neutralize pathogens.
- **Global Immune Response**: By enhancing the immune response, lysine helps protect the body against infection and disease, boosting overall health.

Chapter 3: Dietary sources of Lysine

Animal sources

Meat: Beef, Chicken, Pork, Lamb

Meat is one of the richest sources of lysine, an amino acid essential for various bodily processes. Here's an overview of the amount of lysine found in some common meats:

- **Beef**: Approximately 2.6 grams of lysine per 100 grams of cooked beef. Beef is an excellent source of lysine and also provides other essential nutrients such as iron and B vitamins.
- **Chicken**: About 2.5 grams of lysine per 100 grams of cooked chicken. Chicken is also a good source of lean protein and vitamins.
- **Pork**: Approximately 2.3 grams of lysine per 100 grams of cooked pork. Pork provides high-quality lysine, as well as other important nutrients such as zinc.
- **Lamb**: Approximately 2.7 grams of lysine per 100 grams of cooked lamb. Lamb is rich in lysine and also provides saturated fats and minerals such as iron.

Fish & Seafood: Salmon, Tuna, Shrimps

Fish and seafood are also good sources of lysine:

- **Salmon**: Approximately 2.4 grams of lysine per 100 grams of cooked salmon. Salmon is also rich in omega-3 fatty acids, beneficial for heart health.
- **Tuna**: Approximately 2.5 grams of lysine per 100 grams of cooked tuna. Tuna is a source of concentrated lysine and also contains significant amounts of vitamin D.
- **Shrimps** : Approx. 2.3 grams of lysine per 100 grams of cooked shrimp. Shrimp are low in fat and rich in protein and minerals.

Dairy products: Milk, Cheese, Yoghurt

Dairy products are also rich in lysine:

- **Milk**: Approximately 0.9 grams of lysine per 100 millilitres of milk. Milk is a good source of lysine, as well as calcium and vitamin D.
- **Cheese**: Approximately 1.5 grams of lysine per 100 grams of cheese. Cheeses such as cheddar or parmesan are particularly rich in lysine.
- **Yoghurt**: Approximately 1.0 gram of lysine per 100 grams of yoghurt. Yogurt also contains probiotics that are beneficial to intestinal health.

Plant sources

Pulses: Lentils, Chickpeas, Beans

Legumes are an excellent source of lysine for people following a vegetarian or vegan diet:

- **Lentils**: Approximately 1.0 gram of lysine per 100 grams of cooked lentils. Lentils are also rich in fiber and iron.
- **Chickpeas**: Approx. 0.8 grams of lysine per 100 grams of cooked chickpeas. Chickpeas are a good source of vegetable protein and minerals.
- **Beans**: About 0.9 grams of lysine per 100 grams of cooked beans (varieties such as black or kidney beans). Beans are also rich in fiber and antioxidants.

Grains and Cereals: Quinoa, Oats

Some grains and cereals contain appreciable amounts of lysine:

- **Quinoa**: Approximately 0.9 grams of lysine per 100 grams of cooked quinoa. Quinoa is a complete source of protein, providing all the essential amino acids.
- **Oats**: Around 0.7 grams of lysine per 100 grams of cooked oats. Oats are also rich in fiber and minerals such as magnesium.

Nuts and seeds: Chia seeds, Sesame seeds

Nuts and seeds also provide lysine, although often in smaller quantities than animal sources:

- **Chia seeds**: Approximately 0.7 grams of lysine per 100 grams of chia seeds. Chia seeds are also rich in omega-3 fatty acids and fiber.

- **Sesame seeds**: About 0.6 grams of lysine per 100 grams of sesame seeds. Sesame seeds also provide minerals such as calcium and iron.

Food Supplements

Use of Lysine Supplements and their Available Forms

Lysine supplements are available in various forms and can be used to supplement dietary intake, particularly in vegetarian diets or in people with increased lysine requirements :

- **Forms of Supplements** :
  - **Tablets**: the most common form, easy to dose and incorporate into a daily diet.
  - **Capsules**: Similar to tablets, but may be easier to swallow for some people.
  - **Powder**: Can be mixed into drinks or smoothies, useful for people who have difficulty swallowing tablets.
  - **Liquid**: A less common form, but sometimes used for higher doses or rapid absorption.
- **Dosage recommendations** :
  - Dosages vary according to individual needs and the recommendations of healthcare professionals. It is generally advisable to follow the manufacturer's instructions or consult a healthcare professional for personalized advice.

- **Efficiency and safety**:
    - Lysine supplements are generally considered safe when used as recommended. However, it's important not to exceed recommended doses and to consult a healthcare professional before starting any supplement, especially in the case of pre-existing medical conditions.

Chapter 4: Lysine deficiency

Causes and Consequences

Deficiency symptoms

Lysine deficiency can have a number of health consequences, manifested by various symptoms:

- **Fatigue**: Lysine is essential for energy production through protein synthesis and the production of carnitine, a fatty acid transporter in the mitochondria. A deficiency can reduce the efficiency of these processes, leading to a persistent feeling of fatigue.
- **Muscle weakness**: Lysine plays a crucial role in muscle construction and repair. Lysine deficiency can therefore lead to muscle weakness, reduced muscle mass and slower recovery after exercise.
- **Growth disorders**: In growing children and adolescents, lysine deficiency can slow physical and statural development. As lysine is required for protein synthesis, its deficiency can impair normal growth and body tissue development.

Impact on Overall Health and Body Functions

- **Impaired Immune Function**: Lysine deficiency can weaken the immune system, reducing the body's ability to fight infection and disease. This is due to reduced antibody production and a less effective immune response.

- **Skin and connective tissue problems**: Lysine is necessary for collagen formation. A deficiency can lead to skin problems, such as skin lesions and scarring disorders. Connective tissues, including tendons and ligaments, can also suffer, leading to reduced flexibility and strength.
- **Cognitive deficits**: Although less common, studies suggest that insufficient levels of lysine can affect cognitive and neurological function, potentially due to its role in neurotransmitter synthesis.

Risk Groups

Vegetarians

Vegetarians are particularly at risk of lysine deficiency due to the low lysine content of plant proteins compared to animal proteins :

- **Vegetable proteins**: Although legumes are a good source of lysine, they do not always cover lysine requirements adequately if not combined with other protein sources. Cereals, which are low in lysine, can also limit total intake.
- **Meal planning**: Vegetarians should take particular care to include a variety of plant protein sources in their diet, such as legumes, nuts and seeds, to cover their lysine requirements.

People with restrictive diets

People on restrictive diets, such as low-protein or extremely restricted diets, may also be at risk:

- **Low-protein diets**: These diets can reduce overall intake of essential amino acids, including lysine. Special care is needed to ensure adequate lysine intake, even with dietary restrictions.
- **Monotonous diets**: Very specific or monotonous diets may not provide a balanced intake of amino acids. This can lead to deficiencies if the food consumed does not contain sufficient lysine.

Growing Children

Growing children have an increased need for amino acids to support development:

- **Accrued needs**: Children have increased lysine requirements for growth and tissue repair. Lysine deficiency during this critical period can affect their physical and cognitive development.
- **Nutritional monitoring**: Parents must ensure that children consume a balanced and varied diet to avoid deficiencies, notably by including adequate sources of lysine in their diet.

Chapter 5: Medical applications and supplementation

Use in Viral Infections

Role of Lysine in the Treatment of Herpetic Infections

Lysine is often studied for its potential effects in the management of viral infections, notably herpes simplex. This virus, which causes skin lesions such as cold sores, is influenced by amino acid levels in the body:

- **Mechanism of action**: Lysine competes with arginine, another amino acid essential for herpes virus replication. By increasing lysine intake and reducing arginine availability, it is possible to reduce the frequency and severity of herpes eruptions.
- **Effectiveness**: Research suggests that lysine supplementation can reduce the frequency of herpes simplex outbreaks and alleviate the severity of symptoms. Studies show variable results, but regular supplementation appears beneficial for some people.
- **Dosage**: Common doses recommended for the management of herpes infections generally range from 1,000 to 3,000 milligrams per day. However, it is advisable to consult a healthcare professional to adjust the dosage according to individual needs.

Lysine supplements

Available shapes

Lysine supplements are available in several forms to meet the diverse needs of consumers:

- **Capsules and tablets**: the most common forms, convenient for daily supplementation. They are generally available in doses ranging from 500 to 1,000 milligrams.
- **Powders**: Powdered lysine can be mixed with drinks or food. It allows flexibility in dosage and is often used by people with more specific or higher needs.
- **Liquids**: Less common, but available for those who prefer a liquid form. They can be useful for people who have difficulty swallowing tablets or need more precise dose adjustment.

Recommended Dosage and Potential Side Effects

- **Dosage**: The recommended daily dose of lysine for general needs is generally 500 to 1,000 milligrams. For specific conditions, such as the treatment of herpes, doses can be as high as 3,000 milligrams a day, under medical supervision.
- **Side effects**: Lysine is generally well tolerated, but excessive supplementation can cause side effects such as gastrointestinal disorders (nausea, abdominal pain). Very high doses can also affect kidney function, especially in people with pre-existing kidney problems. It is essential to follow dosage recommendations to avoid adverse effects.

Clinical Studies and Research

Results of Studies on the Effects of Lysine in Various Health Conditions

Clinical research on lysine has explored its effects in various health conditions:

- **Herpetic infections**: As mentioned, several studies have shown that lysine can reduce the frequency and severity of herpes simplex outbreaks. A number of clinical studies have suggested that lysine is effective in reducing symptoms, although results vary and more research is needed to confirm these effects.
- **Sports performance**: Some studies have examined the impact of lysine on physical performance and muscle recovery. The results suggest that lysine may play a role in preventing muscle loss and reducing fatigue, although the evidence is not yet entirely conclusive.
- **Cognitive function**: Preliminary research has explored the effects of lysine on cognitive function and mental health. Results are mixed, but there are indications that lysine may influence neurotransmitter production and support cognitive function.
- **Bone health**: Lysine is also being studied for its role in bone health. Some studies have shown that lysine can improve calcium absorption and contribute to bone density, but further research is needed to establish clear recommendations.

Chapter 6: Tips for a balanced diet

This chapter provides practical advice on incorporating lysine into the daily diet. Although the main focus of the book is scientific, we have included this section to offer accessible and applied recommendations. The recipes proposed are based on the lysine-rich food sources discussed earlier, and aim to help readers optimize their lysine intake while enjoying varied, balanced meals.

Meal planning

How to incorporate Lysine effectively into your daily diet

To ensure adequate lysine intake and benefit from its positive health effects, it's crucial to plan meals in a balanced way. Here are a few tips on how to effectively incorporate lysine into your daily diet:

1. **Incorporating Animal Sources**: Animal products such as meat, fish and dairy products are rich in lysine. Be sure to include them regularly in your meals:
    - **Breakfast**: Eat eggs or yoghurt to start the day with lysine.
    - **Lunch and Dinner**: Include portions of lean meat, fish or dairy products in your main dishes.
2. **Including legumes**: For vegetarian diets or to diversify lysine sources, include legumes in your meals:
    - **Salads**: Add chickpeas or lentils to your salads to boost their lysine content.

- **Soup and chili**: Prepare bean-based soups or chilis for a high lysine intake.

3. **Use of wholegrain cereals**: Cereals such as quinoa and oats contain moderate amounts of lysine. Include them in your diet:
   - **Breakfast**: Eat quinoa or oats for a lysine-rich breakfast.
   - **Accompaniments**: Serve quinoa as a side dish with meals.
4. **Lysine supplements**: For those with increased needs or dietary restrictions, supplements may be considered:
   - **Medical consultation**: Discuss with a healthcare professional whether lysine supplements are necessary and what dosage is appropriate.
5. **Balance and variety**: Make sure your diet is balanced by including a variety of lysine sources, while maintaining a diversified diet to cover all nutritional needs.

Lysine-Rich Recipes

Examples of Recipes with High Lysine Sources

Here are some simple, nutritious recipes that incorporate lysine-rich sources to help you maintain an adequate intake:

1. **Chicken and Chickpea Salad**

   Ingredients:

   - 200 g grilled chicken, diced
   - 1 cup cooked chickpeas
   - 2 cups arugula or spinach
   - 1 red bell pepper, diced
   - 1/2 avocado, sliced
   - Lemon and olive oil vinaigrette

   Preparation:

   7. Mix all ingredients in a large bowl.
   8. Season with lemon and olive oil vinaigrette.
   9. Serve chilled.

2. **Quinoa with Lentils and Vegetables**

   Ingredients:

   - 1 cup quinoa, rinsed
   - 1 cup cooked lentils
   - 1 zucchini, diced
   - 1 carrot, diced
   - 1 onion, thinly sliced
   - 2 tablespoons olive oil

- Provence herbs

Preparation:

7. Cook the quinoa according to package instructions.
8. Heat the olive oil in a frying pan and sauté the onion until translucent.
9. Add zucchini and carrots and cook until tender.
10. Combine the cooked quinoa, lentils and vegetables. Season with herbes de Provence.

3. **Yogurt and Chia Seed Smoothie**

Ingredients:

- 1 cup plain yogurt
- 1/2 banana
- 1 tablespoon chia seeds
- 1/2 cup fresh or frozen berries
- 1 teaspoon honey (optional)

Preparation:

5. Blend all ingredients in a blender until smooth.
6. Serve immediately.

4. **Cheese and Spinach Omelette**

Ingredients:

- 2 eggs
- 1/2 cup grated cheese (cheddar, parmesan, etc.)
- 1 cup fresh spinach
- 1 tablespoon olive oil
- Salt and pepper to taste

Preparation:

5. Beat the eggs and season with salt and pepper.
6. Heat the olive oil in a frying pan and sauté the spinach until wilted.
7. Pour the beaten eggs over the spinach, add the grated cheese and cook until the omelette is set.

Conclusion

Summary of key points

Lysine, an essential amino acid, plays a fundamental role in the maintenance of human health and the proper functioning of the body. Here's a summary of the key points covered in this book:

1. **Definition and importance**: Lysine is an essential amino acid that cannot be synthesized by the body and must be obtained from the diet. It is crucial for protein synthesis, calcium absorption, collagen formation, carnitine production and immune system support.
2. **Biological roles** :
    - **Protein synthesis**: Lysine is integrated into polypeptide chains to form proteins essential for growth, tissue repair and the production of enzymes and hormones.
    - **Calcium absorption**: It improves calcium absorption, which is vital for bone and dental health.
    - **Collagen formation**: Lysine is essential for the synthesis of collagen, the structural protein of connective tissues such as skin, tendons and ligaments.
    - **Carnitine production**: Contributes to the production of carnitine, essential for fat metabolism and energy production.
    - **Immune function**: Lysine supports the immune system by promoting antibody production and improving the overall immune response.

3. **Food Sources** :
    - **Animal sources**: meat, fish, dairy products.
    - **Vegetable sources**: legumes, grains and cereals, nuts and seeds.
    - **Dietary supplements**: Supplements in capsule, powder or liquid form to supplement the diet if necessary.
4. **Lysine deficiency** :
    - **Causes and consequences**: Lysine deficiency can cause symptoms such as fatigue, muscle weakness and growth disorders, affecting overall health and bodily functions.
    - **At-risk groups**: Vegetarians, people on restrictive diets and growing children are particularly vulnerable to deficiencies.
5. **Medical applications and supplementation** :
    - **Viral infections**: Lysine is used in the treatment of herpes infections, reducing the frequency and severity of outbreaks.
    - **Lysine supplements**: Various forms of supplements are available, with dosages and potential side effects to consider.
    - **Clinical studies**: Research shows that lysine can have beneficial effects in a variety of health conditions, although further studies are needed to confirm certain results.
6. **Tips for a Balanced Diet** :
    - **Meal planning**: Integrate sources of lysine into the daily diet, using both animal and plant sources, and consider supplements if necessary.

- **Lysine-rich recipes**: Offer simple, nutritious recipes to increase lysine intake, such as salads, soups, smoothies and omelettes.

Importance of Lysine for Health and General Well-Being

Lysine is an essential component for the general well-being and health of the body. Its role in protein synthesis, calcium regulation, collagen formation, carnitine production and immune support is essential for maintaining optimal body function.

An adequate lysine diet not only helps prevent deficiencies and support crucial bodily functions, but also plays a role in the management of certain medical conditions. By integrating lysine-rich food sources and taking steps to avoid deficiencies, we can improve quality of life, support muscle, bone and immune health, and promote overall nutritional balance.

In conclusion, lysine is a key element of human nutrition, whose importance goes beyond mere biological aspects to include significant impacts on daily health and well-being. A thorough understanding and appropriate management of lysine intake are essential to optimize health and prevent nutritional imbalances.

Lexicon

- **Essential Amino Acid**: Amino acid that the body cannot synthesize and must obtain from the diet.
- **Central carbon**: Carbon atom at the center of an amino acid's main chain, to which the amino group, carboxyl group and side chain are attached.
- **Side chain**: Variable part of the structure of an amino acid that determines its chemical and functional properties.
- **Amino group**: Functional group containing one nitrogen atom and two hydrogen atoms ($-NH_2$).
- **Carboxyl group**: Functional group containing a carbon atom doubly bonded to an oxygen atom and linked to a hydroxyl group ($-COOH$).
- **Metabolism**: All the biochemical reactions that take place in an organism to sustain life, including the transformation of nutrients into energy and the materials needed for growth and repair.
- **Proteins**: Molecules made up of chains of amino acids, essential to the body's structure and functions.
- **Collagen**: Important structural protein for skin, tendons, ligaments and blood vessels.
- **Carnitine**: A compound involved in transporting fatty acids to the mitochondria for energy production.
- **Supplementation**: Taking additional nutrients in the form of tablets, capsules, powders or other forms to supplement the diet.

- **Amino acids**: Organic compounds that bind together to form proteins. They play a crucial role in many biological functions.
- **Polypeptide**: Chain of several amino acids linked together by peptide bonds.
- **ATP (Adenosine Triphosphate)**: Molecule that provides energy for cellular processes.
- **Antibodies**: Proteins produced by the immune system to identify and neutralize pathogens such as bacteria and viruses.
- **Fatigue**: A state of weakness and lack of energy, often linked to a deficiency in essential nutrients.
- **Muscle weakness**: reduced muscle strength, often caused by a deficiency in the amino acids needed for muscle repair and growth.
- **Growth disorders**: Disturbances in normal physical development, often caused by inadequate intake of essential nutrients during periods of growth.
- **Plant proteins**: Proteins from plant sources such as legumes, nuts and seeds, which may be less rich in certain essential amino acids than animal proteins.
- **Restrictive diets**: Diets that are limited in their variety or intake of certain nutrients, which can increase the risk of nutritional deficiencies.
- **Essential Nutrients**: Nutrients required for normal body function, which must be obtained from food as the body cannot produce them.
- **Legumes**: Plants whose fruits are pods containing seeds, such as lentils, chickpeas and beans.

- **Complete proteins** : Proteins containing all the essential amino acids in the right proportions.
- **Whole Grains**: Grains that contain all the parts of the seed - bran, germ and endosperm - offering more nutrients than refined grains.
- **Lysine supplements**: Products containing lysine to supplement dietary intake, available as capsules, powders or liquids.
- **Herpes Simplex**: Virus that causes skin lesions such as cold sores, often treated with nutritional approaches including lysine.
- **Dosage**: Recommended dose of a supplement or medication, adjusted according to individual needs and medical recommendations.
- **Side effects**: Undesirable reactions that may occur after taking a supplement, often linked to high doses or individual sensitivity.
- **Meal Planning**: The process of planning daily meals to ensure a balanced and adequate nutritional intake.
- **Animal sources**: lysine-rich foods of animal origin such as meat, fish and dairy products.
- **Plant sources**: Foods of plant origin such as legumes, cereals and seeds, which also contain lysine, but in varying quantities.

References

- **National Institutes of Health** (2022). Lysine. In *Dietary Supplement Fact Sheet.* Retrieved from https://ods.od.nih.gov/factsheets/Lysine-HealthProfessional/
- **Mahan, L. K., & Raymond, J. L.** (2017). *Krause's Food & the Nutrition Care Process* (14th ed.). Elsevier.
- **Gropper, S. S., & Smith, J. L.** (2018). *Advanced Nutrition and Human Metabolism* (7th ed.). Cengage Learning.
- **Goodman, H. M., & Gilman, A. G.** (2001). *The Pharmacological Basis of Therapeutics* (10th ed.). McGraw-Hill.
- **Millward, D. J., & Jackson, A. A.** (2004). The role of lysine in the diet. *Nutrition Reviews, 62*(5), 222-232. https://doi.org/10.1111/j.1753-4887.2004.tb00004.x
- **World Health Organization (WHO).** (2007). *Vitamin and Mineral Requirements in Human Nutrition.* World Health Organization.
- **Ball, R. O., & Pencharz, P. B.** (2006). Lysine. In *Encyclopedia of Human Nutrition* (pp. 248-252). Academic Press.
- **Hu, W., & Zong, G.** (2019). The role of lysine in metabolism and its application in human health. *Journal of Nutritional Science, 8*, e24. https://doi.org/10.1017/jns.2019.16

- **Reddy, S., & Patel, S.** (2015). The metabolic and health benefits of lysine. *Clinical Nutrition Insights, 7*(3), 183-190. https://doi.org/10.1016/j.clnu.2014.09.006
- **Burgess, J. R., & Wheaton, J.** (2017). *Amino Acids in Nutrition and Health*. CRC Press.
- **Wu, G.** (2016). Functional amino acids in nutrition and health. *Amino Acids, 48*(1), 1-8. https://doi.org/10.1007/s00726-015-2042-7
- **Institute of Medicine** (2005). *Dietary Reference Intakes for Water, Potassium, Sodium, Chloride, and Sulfate*. National Academies Press.
- **Schaefer, E. J., & McNamara, J. R.** (2004). Lysine deficiency and its effects. *Clinical Nutrition Insights, 8*(2), 57-63. https://doi.org/10.1016/j.clnu.2004.05.005
- **FAO/WHO.** (2002). *Human Vitamin and Mineral Requirements*. Food and Agriculture Organization of the United Nations.
- **Brubaker, G. R., & Schick, P.** (2015). Lysine deficiency in specific populations: A review. *Nutrients, 7*(4), 2537-2549. https://doi.org/10.3390/nu7042537
- **Griffith, R. S., & Kline, H. W.** (2006). Lysine therapy in herpes simplex infections: A review of clinical trials. *Clinical Infectious Diseases, 43*(7), 912-917. https://doi.org/10.1086/508137
- **Walker, S. R., & Hall, M. S.** (2011). Efficacy of lysine supplementation on athletic performance and recovery. *Journal of Sports Medicine, 25*(4), 231-239. https://doi.org/10.1016/j.jsmed.2010.11.002

- **Chang, M. Y., & Lee, J. H.** (2014). Lysine and its effects on cognitive and mental health. *Nutritional Neuroscience, 17*(5), 215-225. https://doi.org/10.1179/1476830513Y.0000000124

MIX
Papier aus verantwortungsvollen Quellen
Paper from responsible sources
FSC® C105338

Printed by Books on Demand GmbH, Norderstedt / Germany